There's a Fortune in Worms

The Plough is one of the most ancient and most valuable of man's inventions; but long before he existed the land was in fact regularly ploughed and still continues to be thus ploughed by earthworms. It may be doubted whether there are many other animals which have played so important a part in the history of the world as have these lowly organized creatures.

CHARLES DARWIN

There's a Fortune in WORMS

Hank Haynes

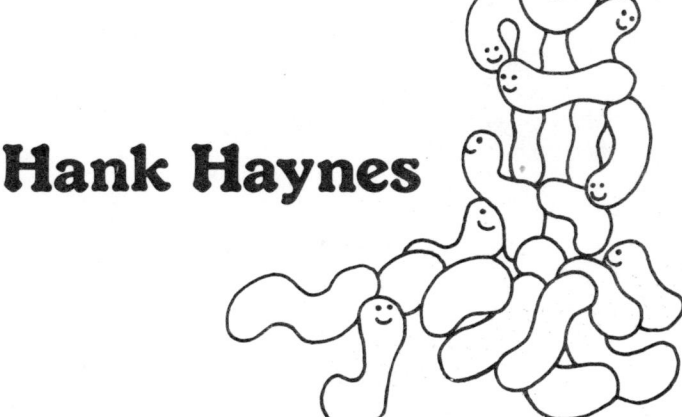

Brooke House   Los Angeles, California

Copyright © 1976 by Hank Haynes

All rights reserved. This book may not be reproduced in whole or in part by any means, whether mechanical or electronic, without the written permission of Brooke House Publishers, Northridge, California 91324, except for brief passages used in a review. Manufactured in the United States of America

Revised and enlarged edition

Library of Congress Cataloging in Publication Data
Haynes, Hank. There's a fortune in worms.
1. Earthworm culture. 2. Earthworms—Marketing. I. Title.
SF597.E3H39   1976   338.3'71'7546   76-7316
ISBN 0-912588-18-7

67890123987654321

Design and illustration by Barbara Monahan
Photography by Ron Sheridan and Bill Courtice
Quotation by Henry Hopp on page 2 reprinted with permission of Garden Way Publishing

Contents

Introduction 3
Kinds of Worms To Grow 7
Getting Started in Worm Ranching 8
Bed Size for Worms 14
Worm Care 16
Worm Enemies 20
Harvesting Worms 23
Splitting Beds 25
Worm Castings 27
Worms and the Food Crisis 29
Worm Reproduction 31
Selling Worms 33
Other Ways To Sell Worms 38
Helping Others Get into the Worm Business 40
Growing Worms in an Apartment 43
When To Begin Selling Worms 46
Cost of Getting Started in Worm Ranching 48
Raising African Night Crawlers 50
Conclusion 52

There's a Fortune in Worms

The association of earthworms with productive soil causes people to wonder if earthworms play a positive role in soil productivity or merely prefer to live in the better soils without contributing to their productivity. This is not just an academic question. It is one that can vitally affect the productivity of many millions of acres of agricultural soil where farming or gardening is being carried on without regard to what it may be doing to the earthworm population.

The newest researches have amply demonstrated that earthworms do have very important effects on the productivity of soil. They are by no means just passive denizens of the soil, without effect on its properties. Quite the contrary, controlled experiments now show that some of the important properties of certain kinds of soil are directly attributable to the activity of the earthworms; and that when the earthworms are absent these properties are altered.

DR. HENRY HOPP
U.S. Department of Agriculture

Introduction

Perhaps your interest in earthworms is due to that wonderful thing called curiosity. However, if you knew the facts, your curiosity would suddenly vanish; and armed with the right kind of knowledge, you would most likely take immediate steps to get started in the worm business. That is the purpose of this book—to introduce you to what could become one of the most profitable businesses in America.

It is incredible that more people are not growing worms for a living, especially when you consider that millions of worms are grown and shipped every year and yet the supply never meets the demand. There are never enough worms to go around.

Now, how can this be? Is the thought of growing worms distasteful to people? Does growing worms require hard work and long hours? Does it require a large investment, lots of land, or costly facilities? Is there really money to be made?

Unlike the milk and farming industries which are highly publicized, the earthworm industry is virtually unknown to the public. Worm growers have not bothered to form associations or co-ops. They do not employ lobbyists to represent their best interests, operate from big buildings with neatly landscaped surroundings, employ prestigious, national advertising agencies, nor give away fancy, multi-colored catalogs.

Worm growers are typical citizens, quietly going about the business of raising worms in their backyards or on small ranches. They are not trained in the world of big business, nor do they care about big-business ideas. They generally follow a similar, simple, business pattern. They grow their worms and sell them through small ads placed in the back pages of sporting or general purpose magazines. After they receive their orders, they pack the worms in special containers, which they then mail to customers around the country. There are hundreds of growers, men and women alike, doing exactly as described, operating privately and independently from one another and collecting millions of dollars every year.

Over the years, the primary market for earthworms has been fishermen, fish hatcheries, and bait dealers. The fishing industry grows every year, and will continue to grow, but it is fast becoming a secondary market for worm sales. The new, booming market consists of some 30 million home gardeners and thousands of organic farmers who have suddenly discovered that earthworms improve soil. If you have any doubts or reservations about the marketability of worms, you should read the absorbing report contributed by Dr. Henry Hopp of the U.S. Department of Agriculture on earthworms and their effect on soil and plant life. The agricultural market, as you will learn later, is so overwhelming that it is questionable if we will ever be able to grow enough worms to satisfy the demand.

My ranch, World Wide Worms, Inc., operates similar to hundreds of worm ranches or backyards, but perhaps it is larger than most. Let me give you a thumbnail sketch of my operation because I can relate my experiences from firsthand knowledge. To begin with, I started my worm ranch with 5 beds, each containing about 100,000 worms. My total investment was less than $2,000. One year later, my 5 beds had grown to 80 beds having a conservative value of $32,000. As you will see, the word "conservative" should be emphasized. During the first year of operation, I spent less than one day per week performing watering and feeding chores and building new beds. After months had passed, it dawned on me that I had discovered a fantastic outdoor occupation. However, I was practical enough to realize that having fun was one thing and bringing home the bacon was something entirely different. Since bills were accumulating, I decided that I had better start selling my worms.

Let me illustrate the profit potential. Eighty beds produce about 800 pounds per month without depleting the worm population (worms double in population every 60 to 90 days). The current wholesale price for worms is $2.50 per

Partial view of the author's earthworm ranch.

pound; therefore, 800 pounds times $2.50 equals $2,000 per month or $24,000 per year. But I do not recommend selling worms at wholesale prices. The retail price, believe it or not, averages $8.00 per pound for bed-run (mixed sizes) and $12.00 per pound for hand selected, breeder-size worms. I also know that some ranchers are selling worms for soil improvement for as much as $20.00 per pound. Just for fun, multiply $20.00 times 800 pounds—it's $16,000!

Now you can begin to see why you should be in the worm business.

Kinds of Worms To Grow

There are over 3,000 different species of earthworms. All of the instructions and information covered in this book relate to one species only, an earthworm commonly referred to as a Manure Worm, Red Worm, Red Wiggler, Red Wriggler, and Red Hybrid. There are actually two types of the same basic species: the helodrilus foetidus, a worm that has transverse rings of yellow and maroon color which alternate the length of its body; and the lumbricus rubellus, a worm that has a deep maroon color throughout its entire body. Both types are present together in commercially grown worm beds, but the lumbricus rubellus comprises 90 percent of all worm sales in the United States.

The African Night Crawler (lumbricus terrestris) is also a commercially grown earthworm, but it is raised solely for the fishing industry. African Night Crawlers are not grown for soil improvement because of environmental requirements which inhibit their ability to reproduce.

Getting Started in Worm Ranching

There are eight basic steps to follow in order to get started in worm ranching.

One

Find a level spot somewhere in your backyard. It is not necessary to remove the lawn or weeds unless by doing so you establish a level area.

Two

Out of new or used lumber, build a worm bed 4-foot wide by 8-foot long by 1-foot high without any bottom. The lumber may be 1 inch by 12 inch or 2 inch by 12 inch. The heavier lumber is preferable because of its stability and long life. Also, you can improvise by using whatever lumber you have on hand or by using scrap lumber nailed together. Do not use redwood because it contains an acid harmful to worms. The important thing is that your bed measures 4-foot wide by 8-foot long by 1-foot high without any bottom and resting level on the ground. Then take some loose dirt and deposit it along the inside edge, where the wood touches the ground.

Three

Place bedding material inside the bed about 6-inches deep and spread evenly throughout the bottom. The bedding material can consist of almost any kind of organic matter such as sawdust, wood chips, grass cuttings, peat moss, decomposed manure (horse, cow, steer, or rabbit), hay, alfalfa, or a combination of these materials. Do not use redwood sawdust, redwood chips, or any inorganic material such as dirt. Also, be sure that all organic material used has passed the heating stage. As you probably know, decomposition of organic matter is a process of bacteria and microorganism activity which is stimulated by the introduction of moisture. This activity creates tremendous heat and will surely harm your worms unless they can escape from it. Since worms usually can escape, they will probably crawl out of your bed and disappear into the surrounding garden, thereby leaving you without worms to grow!

The use of chicken manure generally is not recommended as either bedding or feeding material. It can be used satisfactorily if it is properly decomposed and leached of harmful salts and acids. This can be accomplished by spreading the manure over the ground about 12-inches thick and wetting it down thoroughly once each week. The aging, leaching, and decomposition will take about three to four months, but it can be accomplished faster with the introduction of commercial microorganism and bacteria activators. If you plan to use chicken manure in your worm beds, it would be helpful for you to read a good book on composting.

Four

Sprinkle 1 or 2 cups of limestone flour (calcium carbonate) over the entire bedding surface. Limestone flour is inexpensive and is available at any store selling poultry and livestock feed. A 100-pound bag will cost about $2.50 and will last a long time. Do not confuse limestone flour with lye; they are entirely different products. Be sure you buy powdered calcium carbonate. Soak the bedding with water, taking care to wash the limestone into the bedding depths. Limestone will aid in neutralizing acid, will reduce or eliminate odor, and will help keep pests out of your bed.

Five

Place your worms in the bedding by spreading them evenly over the entire bedding surface. Worms are very sensitive to light and will promptly crawl into the depths of the bedding.

Six

The steps taken so far will provide a comfortable, happy, and livable environment for your worms. The bedding material is good food, which the worms will eventually eat up, but continuous feeding is still necessary.

Red Worms by nature are top feeders; that is, they prefer to do their eating at the very top surface of the worm bed. When they are not eating aggressively, their habit is to crawl into the depths of the bed and move about or mate with other worms. Even when the worms are moving about or mating, they still continue to eat, but on a limited scale. In order for you to raise healthy, lively worms, you should continue to provide feed on the top surface as the feed disappears.

Obtain a free source of rabbit, steer, cow, or horse manure. Do not use chicken manure unless it is well decomposed and leached of salts and acids. Be calm about the prospect of using manure and having to go out and locate it. Manure is plentiful and is really not distasteful matter. Your worms will love it and multiply in it, and you will make money with it.

Your first feeding will not require a large amount of manure; a full 4-cubic-foot wheel barrel should be ample. Water the manure thoroughly and soak in about 1 cup of limestone. If you use a wheel barrel, provide a drain hole in the bottom to drain off excess water. A wooden box or a wash tub will do fine for the soaking operation, but be sure to provide drainage to remove excess water containing acids. Mix and turn the manure until all the particles are moist to be certain that the manure is thoroughly, not just partially, soaked. Place the moist manure on top of the bedding material, and spread it out evenly to about 2-inches thick. Keep the feed area about 6-inches away from the edges of the bed; in other words, the feed area should measure about 3 foot by 7 foot.

Another, perhaps simpler, way to prepare your feed is to lay the manure on the ground about 12-inches thick. Sprinkle limestone on top and wet it down thoroughly for two to three hours with a garden sprinkler.

If you get your manure from a horse ranch, be sure to find out if the owner uses a deworming agent on the horses. If the answer is yes, ask what kind of poison is being used. One variety of deworming chemical kills the worms and might kill your worms, too, if any appreciable amount remains in the manure. Other deworming agents are not harmful because they merely tranquilize the worms temporarily.

You should also avoid sacked manure sold at nurseries and garden shops. This bagged manure is expensive and usually treated with harmful chemicals and weed killers. It is also questionable whether sacked manure has any nutritional value.

Seven

Cut an old piece of used carpet so that it measures approximately 3 foot by 7 foot. Place the carpet directly over the feed area with the back of the carpet facing down; that is, the carpet knap should be facing up, not touching the feed area. Do not use rubber-backed carpet because it will not allow for adequate air circulation.

Used carpet is easily found; in fact, some of your local carpet stores might give you all you want for the asking. If this turns out to be a problem for you, substitute an old army blanket, burlap sacks, or newspapers for the carpet. Whatever you use, be sure that water and air can easily pass through it. After you have installed the covering, wet it down well.

The carpet, blanket, burlap bag, or newspaper covering is important to the general goodwill of your worm bed. The covering will help retain moisture at the feeding area, provide a dark environment, and induce the worms to feed at the very top surface of the bed.

Simply constructed bed with hinged plywood cover and without bottom.

You will probably be interested to know that your worms will eventually eat the entire carpet and will turn every bit of it into mineral-rich castings unless the carpet is composed of inorganic, synthetic materials such as nylon. The latter will not interest the worms because they feed only on things that are derived from living matter.

Eight

The last step requires that you cover your bed frame with a lid. Because of its convenience in handling, a 4-foot-by-8-foot plywood sheet is best; you can hinge one end and easily maneuver it out of the way during watering and feeding chores. You could also use scrap lumber for your lid, but do not use any type of cover that resists free air flow such as vinyl plastic.

It is highly desirable, yet not essential, to cover your bed as described. The benefits far outweigh the costs: During heavy rainstorms, your worms will need some kind of mechanical protection because while worms like moisture, they do not like to be flooded. Also, a cover enhances the dark environment and helps maintain proper moisture content in the bed.

Bed Size for Worms

Such a large bed is not necessary at all. The reason I describe the 4-foot-by-8-foot bed is that it conforms to the basic size used by the successful worm growers in the United States. A bed of this size will contain a population of 100,000 to 200,000 worms; and it seems to be the ideal size for maintenance, watering, feeding, and harvesting.

Inside or outside the house, you can start your worm farm in almost any kind of container, in small wooden boxes, wash tubs, halved oil drums, and even old bath tubs. You could also build your worm bed out of cement blocks or bricks or dig a hole in your backyard. All of these containers will do fine as long as you follow the basic instructions in this book and make certain that you provide all containers you use with adequate drainage.

As you gain experience with the feeding and caring of your earthworms, you may decide to abandon some of the instructions in this book. For example, I no longer find it necessary to use carpets or lids on my beds. I have learned that when the top surface of the bedding material is dry, the dried surface acts as a natural insulator and shields against sunlight and evaporation. The worms seem to survive just as well without cover protection, and I manage much better not having to hassle with carpets and plywood covers. As another example, growing worms in boxes or containers is not necessary. You could successfully grow worms in long rows of manure laid on the ground, thus saving on lumber costs. As a beginner, however, I strongly recommend that you follow the conservative trail, taking whatever precautions are necessary to protect your worms. Later on, you will gather enough knowledge and nerve to experiment with different ways of doing things.

NOTE! this

Long rows of earthworm beds which are practical for large operations.

Worm Care

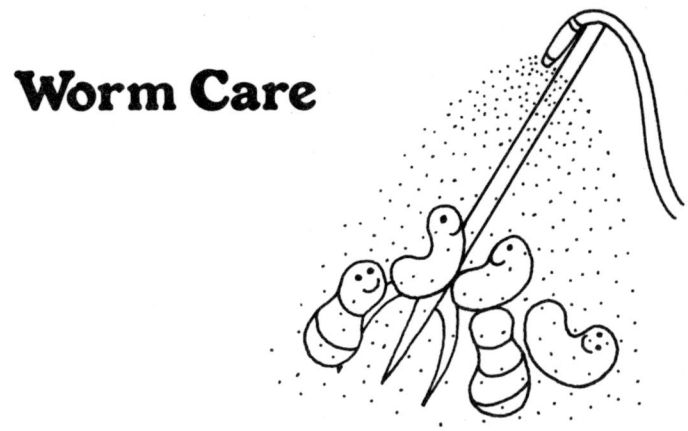

Taking care of your worms will be both easy and fun. It will be very much like planting vegetables in your garden, when you prepare the soil, plant the seeds, irrigate, cultivate, harvest, and eat. The fun and joy of watching nature at work is one of life's most fulfilling experiences. The science of worm culture is very similar to planting a garden: you prepare the bed, plant the worms, water, feed, harvest, and sell.

The importance of keeping your worm bed damp cannot be overstressed. Worms do not eat like other animals. They consume their food by sucking; and obviously, if their food is dry, they cannot ingest. If you let your bed dry out excessively, your worms probably will not starve to death. Instead, their behavior is such that they will most likely crawl out of the bed in their search for a more moist environment in the surrounding area. If this happens, you will not have to worry about watering your beds anymore.

Always keep your bed moist, but not soggy wet. However, there is an exception: some worm growers recommend that you periodically dry out the top feeding surface in order to induce mating. The theory behind this practice is based on the worm's eating and breeding habits. When the mature breeder worm is concentrating on eating, he does not take the time to mate. On the other hand, when the mature breeder worm is concentrating on mating, he is not eating. Therefore, to induce mating on some kind of scientific regularity, one suitable way seems to be to dry the feeding area. This may be advisable practice provided it is not done more frequently than every two months and that the drying out period not exceed four days.

Bed moisture should be checked periodically by visually examining the bedding material. Overwetting is more desirable than underwetting. In either case, do not panic if you are not sure. I have looked at hundreds of beds in various locations and all kinds of conditions. I have seen beds so battered by a bad storm that the beds were left standing in a lake of water and so wet that you could pour the bedding material like water from a glass. I have also seen some terribly neglected beds that had not been watered for several weeks or months and were apparently completely dry. In both of these extremes, the worms were surviving. You have a lot of latitude in controlling the moisture content so do not spend too much of your time worrying about it.

Continually feed your worms as they consume their food. The easiest way to determine if the feed has been consumed is by visual inspection. As worms eat through their food, they ingest the crude organic material and deposit it in the form of castings (excretions). Castings appear to be little mounds of pulverized dirt in a very dark brown, or almost black, color. When this condition is sighted over the entire bed, you have probably waited too long to add fresh feed. But do not be alarmed, there is still food mixed in the depths of the bedding. Your best gauge for adding additional feed is the appearance of patches of pulverized castings measuring 8 to 12 inches in diameter.

If you start worm growing in a small way, many weeks will pass before you will find it necessary to refeed. However, as the worms multiply, grow larger, lay capsules, and hatch eggs, the entire bed will eventually be overrun with worms. When your bed is ready for splitting or harvesting, the worm population will consume about 4 cubic feet of feed in about ten days.

Before placing new feed on your bed, it is recommended practice to pitchfork the entire bed, turning over the bedding material and breaking up the clods. Do not use a shovel because you will damage too many worms. The purpose of turning the bed is very similar to that of cultivating a garden. As you know, when water is regularly applied to soil, the soil tends to compact. In this condition, the flow of water to the root area is restricted. Continual watering of your worm beds will compact the bedding material, too, and will restrict both water flow and worm movements. It is the nature of the worm to move around considerably so he needs loose material. Compacted beds may be corrected by pitchfork-spading the bedding material and gently breaking up the clods. This practice is advisable once a month.

Do regular watering with a spray nozzle. When the weather is cool or damp, watering twice a week should be adequate to keep your bed moist. During the hot summer months, watering once a day should be sufficient. If the weather is extremely hot, that is, over 100 degrees, it may be necessary to sprinkle your bed once in the morning and once in the evening. Be sure that your bed cover is vented slightly during very hot weather in order to ventilate the bed. The moisture content of your bed is important, so test it frequently. Once a week add about 1 cup of limestone, sprinkled evenly over the bed, before watering.

Watering beds is enjoyable on a warm, sunny day, but imagine owning 100 beds that all need to be watered. If you water by hand, figure you will need about three minutes for each bed to water adequately. This means five hours are needed to water 100 beds. As you can see, watering by hand may take too much time. As a solution to this problem, I installed a sprinkler system to help free my time for other things. It is equipped with an automatic sprinkler-control system that turns the water on and off at predetermined times and is adjustable for elapsed time. I am now experimenting with drip irrigation devices that meter water over the entire bed surface. Drip irrigation runs continuously, conserves water, and is trouble free mechanically.

Sprinkler system conveniently mounted on right-hand side of beds.

Worm Enemies

You will be about the only enemy of your worms if you do not water and feed them regularly! You have probably heard of the old expression, "The early bird catches the worm." Well, I suppose birds could be considered an enemy, but a somewhat unlikely one if you follow the instructions in this book. There are some pests, however, of which you should be aware.

Springtails

These are small, snow-white insects measuring about 1/32 inch in length. The body is straight with small feelers on the head. They are not harmful to worms, but they do attack dead worms. Since they feed on the manure and remove valuable nutrients, you should control them. Perhaps the best way to control these pests is to apply limestone heavily on top of the bed and let it dry out for about two days before continuing any further watering. If your bed is overrun with Springtail, you can also eliminate them by applying a thin spray of kerosene over the top of the bed. If not overapplied, kerosene will not harm your worms.

Pinworms

These are small, white worms that multiply rapidly in manure. Commonly called Acid Worms, they live and grow in fresh manure that contains a high content of acid. They will not harm the Red Worm; but if you see a large number of these worms, it indicates that your bed contains excessive acid. The best cure is to liberally apply more limestone.

White and Red Mites

These resemble very small spiders and will not harm your worms.

Centipedes, Grubs, and Sow Bugs

These are natural habitants of the manure environment and will not harm your worms. If you see Sow Bugs in your bed, be happy because they are a good sign that your bed is healthy.

Gophers

These are harmless and will not eat your worms, but they may very likely drive you insane. You must control their activity. When I first began in the worm business, I was going to the ranch every other day to water and to feed worms. I observed a few Gopher mounds located some 50 feet away from my beds. As the days passed by, fresh mounds of new dirt were popping up and working closer to the beds until I became pretty nervous. What would the Gophers do if they got underneath my beds? It was disastrous! Those small, harmless little creatures turned my beds completely upside down: the top layer in my beds ended up as 6 inches of freshly dug earth. Underneath this mess was a mixture of dirt, bedding, and confused worms. Of course, you do not want this to happen; but if it does, use the following remedy. Lay about 6 to 8 inches of new feed on top and water heavily. The worms will quickly work themselves up into the feed area, and the dirt will settle to the bottom.

Moles

These will eat your worms and must be controlled through the use of traps or poison. Fortunately, their travels are readily visible.

Harvesting Worms

To harvest worms, you simply take the worms out of their bed, put them into containers, and sell them for money. Worms are sold by count or weight. In other words, some growers actually count the number of worms they sell; others, and this seems to be the most sensible way, take their worms out and weigh them on a scale. There are two basic methods used to harvest.

Baiting

Mold some brown sugar in patty form and place these patties in three or four locations in the bed just under the feed layer. The worms will be overly attracted by the brown sugar and will swarm around the patties in large numbers. Since the worms will tend to ball up around the patties, you can easily remove them in large quantities.

Pyramiding

This is probably the most common method. To begin, lay a flat board across the worm bed or build a sloping traylike table with a lip overhanging the bed. Take a pitchfork full of bedding and place it on the board or tray. There will be worms throughout this mixture. Because worms cannot withstand light, any worms on the surface will quickly retreat into the middle of the material. With your hands or with a brush, slowly remove the bedding material and place it back into the worm bed. Since the removed material will contain quite a few worm capsules and very small baby worms that are difficult to see, be sure to place the removed bedding material back into the bed. As you continue to remove additional material from the pile of bedding, the worms will continue their retreat away from the light until they eventually form a large ball of squirming worms ready for weighing or counting. As soon as you have weighed or counted the worms, place them in a container with a couple of handfuls of bedding material and premoistened peat moss. When storing the harvested worms for shipment, keep them out of direct sunlight.

Harvesting table with tray overhanging the bed.

Splitting Beds

Splitting beds means that you divide one bed to create two beds. The process of splitting becomes necessary as your worm population expands to the point where overpopulation occurs; that is, their home is no longer big enough. Worms must have adequate room to move about or else they will crawl away into the surrounding area.

When your worm bed is ready for splitting, a definite sign of overcrowding appears: the mature adult worm will migrate to the sides and edges of the bed to make room for the young and baby worms. Therefore, if you see all the small worms in the middle of the bed and all the adult worms along the edges, consider either harvesting or splitting. One other sign worth watching is the rate at which the worms consume their food. When approximately 4 cubic feet of new feed disappears in ten days, it is time for splitting.

With the 4-foot-by-8-foot bed, splitting will not be necessary until you attain a population of about 200,000 worms. Once you start splitting, you will find it necessary to continue splitting every 60 to 90 days.

To split beds, begin with building another bed as instructed earlier. Prepare new bedding material. Remove one-half of the contents of your existing bed and put it into your new bed on top of the new bedding material. To remove the contents from the existing bed, start from the middle and work toward the sides using a pitchfork. As you remove the contents from the existing bed and put it into the new bed, you will be removing half of the worms, capsules, feed, and castings. When this step is accomplished, spread the contents evenly throughout the bottom in both beds. Feed, apply limestone, water, and install the carpet covering. That is all there is to it until the next splitting cycle occurs in two to three months.

Worm Castings

Worms feed and live on organic matter. They eat an enormous amount of crude organic matter, in fact, an amount equivalent to their weight every 24 hours. Their excretions are called castings. The worms do a fantastic job of converting crude organic material into castings containing over 5 times the nitrate, 7 times the available phosphorus, 3 times the exchangeable magnesium, 11 times the potash, and 1½ the calcium found in the best top soil in the United States.

Until recently, worm castings have not been considered valuable to worm growers. Now, however, the castings may become more valuable than the worms. Of course, you have to grow worms in order to produce castings. Worm castings are now being nationally advertised and sold as a very rich, pure organic fertilizer. The current retail price of castings is $1.50 per pound. The standard 4-foot-by-8-foot worm bed will produce an annual harvest of 600 pounds of castings. Ten beds could produce a retail value in castings of $9,000 per year; this does not include the sale of any worms. Because there is a worldwide fertilizer problem, do not throw away your castings.

Eventually you will want to clean out your beds to remove excessive buildup of castings. Do this about once a year when your bed is so full of castings that you do not have any space left for feeding. To clean out your bed, first remove all of the worms. After you have removed the worms, take the castings and pile them on the ground in the shape of a pyramid. The castings will contain many egg capsules and baby worms that you will not be able to see.

Water the mound of castings for about two months. Keep a small amount of feed on one end of the mound, and water the feed area to keep it moist. It is also a good idea to cover the small feed area with a piece of carpet. During the following month or so, the capsules will hatch and the young worms will instinctively migrate to the small, but isolated, feed area. In two to three months you can harvest most of the remaining worms from the castings.

Worms and the Food Crisis

There is, indeed, a food crisis. People in the world you live in are starving to death. There are now over 4 billion people occupying this earth and growing at a net population rate of 80 million people per year. In India alone, there are more than 56,000 new mouths to feed every day.

The population explosion and the reduction of the death rate are the major contributors to the food crisis, but there are other factors as well. The worldwide fishing industry, for example, is out of control. Huge fishing fleets harvest fish from the ocean faster than the fish can reproduce. The industry is international, and the governments around the world cannot seem to agree on policy. Without controls, the fishing fleets go blindly ahead, ignoring the future of a starving world.

Another contributor to the food crisis is the increasing cost of petroleum and petroleum by-products used in the manufacture of fertilizers, herbicides, and pesticides. The United States is struggling with oil problems and working on solutions, but countries like India are paralyzed. India's annual food production is not increasing because it is unable to afford the extravagant increases by the oil-producing nations.

Even in the United States, the food crisis is underway. Food surpluses, for example, are a thing of the past. Countries like Russia and China are buying our surplus production at an alarming rate. When you have demand exceeding supply, you have excessive inflation, as evidenced by recent grocery bills.

Where are we going to get the food to feed a starving world? Let us consider the earthworm and examine nature's role for him. His job is to convert anything that once lived, that is, organic, into mineral-rich fertilizer known as castings. He will eat anything as long as it is dead, organic waste. This includes the multitude of things that we wash down our garbage disposal or haul to the dump. Earthworms can convert this waste into top soil, soil in which plants and food can grow.

Red Worms are prolific breeders. They are almost pure protein and a natural food source that can be consumed by the human being. Eating worms may not appeal to you, but your attitude might be different if you were starving to death, like millions are in this world. More and more experimentation is being done to convert earthworms into edible, tasty forms. Earthworms are being dried, ground, and made into almost any shape desired. With the introduction of artificial flavoring, earthworms could be made very palatable.

I visualize worm factories around the world consuming millions of tons of waste, creating millions of tons of pure organic fertilizer, and producing millions of tons of edible earthworms. Does this seem like a fantasy? Well, I believe this to be a realistic prediction and the future of the earthworm business to be utterly fantastic.

Worm Reproduction

Worms are hermaphroditic, that is, each worm has both male and female sexual organs; therefore, each worm has the ability to bear its own family. Mature worms are known as breeder worms and are easily distinguished by the collar or band (clitelum), which is located about one-third of its body down from the tip of the head.

Worms reproduce amazing numbers of offspring. Each breeder will produce one egg capsule every 7 to 10 days. Each capsule contains anywhere from 2 to 20 baby worms, with an average of 4 worms per capsule. Hatching time is 14 to 21 days, and newly hatched worms will reach maturity (breeder size) in 60 to 90 days.

Typically, beds contain worms at all stages of growth: capsules, baby worms, medium-size worms, and mature breeder-size worms. Worms in all stages of growth are called "bed run." Provided they are properly cared for and left undisturbed, worms will double in population every 60 to 90 days. Therefore, a bed containing 100,000 worms (bed run) will grow to a population of 200,000 within 60 to 90 days. When beds reach a population of 200,000 worms, they should be split into two beds to avoid disrupting the growth cycle. As mentioned before, overcrowding of worms can result in loss of worms due to the nature of the mature worms to make room for their babies.

It is important to mention that the growth cycle tends to vary throughout the year. Based on my own experiences, worms multiply faster during the late winter and spring months and slower during the extreme hot and cold seasons. A pretty good, year-round average seems to be that worm beds double in population about every 75 days. Also, worm beds do not always reproduce exactly alike. This may be due to our inability to feed and water each bed precisely the same or due to our inaccuracies in splitting beds. After the splitting operation, it is quite possible to end up with 125,000 worms in one bed and 75,000 in another.

The following chart of growth and reproduction possibilities is purely for illustration. The figures represent a 2½-year history of one bed containing 100,000 worms, doubling in population every 75 days.

Month	*Number of Beds*	*Population*
May	1	100,000
July	2	200,000
October	4	400,000
December	8	800,000
March	16	1,600,000
May	32	3,200,000
August	64	6,400,000
October	128	12,800,000
January	256	25,600,000
March	512	51,200,000
June	1024	102,400,000
August	2048	204,800,000
November	4096	409,600,000

Selling Worms

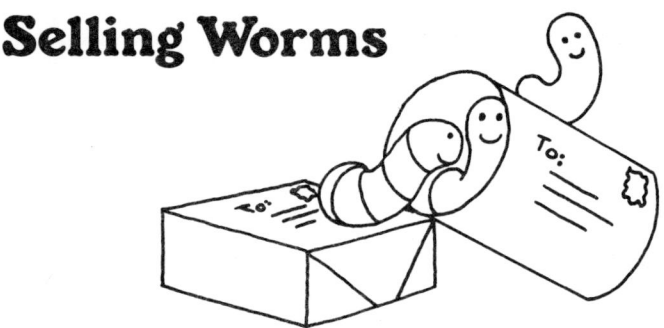

During the course of a year, hundreds of visitors drop by to see what a real worm ranch looks like. Most of my visitors are hungry for information so I end up answering many questions about feeding, watering, reproduction, harvesting, splitting, and income potential. Just when I feel safe, having satisfactorily answered everything put to me, my visitor invariably takes on a suspicious look and says, "Well, you certainly make this business sound easy and profitable; but if I decide to go into it, how will I sell my worms?"

Let us explore that. If you grow worms, do not attempt to sell any until you have developed a good, solid stock because it would be very embarrassing if you received orders for 500,000 worms yet could only deliver 100,000. It would mean writing letters and making excuses and, more importantly, it could cost you the loss of long-term customers, those who would buy from you month after month, year after year. In other words, do not ruin your opportunity just to satisfy your anxiousness to begin selling worms.

While learning the ropes and waiting for your stock to multiply, it would be advisable to contact other worm growers. Explain that you are a fellow grower and that you would like information on price and availability of their worms. These contacts could prove to be valuable because it may become necessary at times to buy worms from them in order to fill your orders. During the peak season, which is late spring to early fall, the demand for worms is so great that widespread shortages are commonplace; therefore, it is in your best interest to develop sources other than your own worms.

All goods and services have to be sold, and selling involves a combination of both persuasion and exposure. For illustration, suppose your local meat market decided to sell top sirloin at ten cents a pound and advertised this great bargain in the newspaper. How long do you think his supply would last? You begin to see how persuasive a good bargain is and how it works to force immediate action. But what if the meat market decided not to advertise, relying only on people drifting in and being told verbally of the special price? Obviously, no persuasion, no exposure, no sale!

Expose your worm business to everyone with whom you come in contact. Tell everybody and pass out business cards to neighbors and at stores, service stations, and social gatherings. The amount of interest and business you will obtain through referrals is surprising.

Prior to making firm marketing plans, choose a simple name for your business. Have your printer help you prepare business letterheads and envelopes. If you do not have a typewriter available, be certain that your correspondence is carefully handwritten, neat, and legible. Send letters requesting rate information for classified advertising to all of the magazines listed below. Explain that you are considering placing an ad in their magazine, and ask for a complimentary copy. They will send you all of the necessary information, including effective sales literature.

Field and Stream, 383 Madison Avenue, New York City, New York 10017

Fur-Fish-Game, 2878 East Main Street, Columbus, Ohio 43209

Organic Gardening & Farming, 33 East Minor Street, Emmaus, Pennsylvania 18049

Outdoor Life, 380 Madison Avenue, New York City, New York 10017

Popular Mechanics, 224 West 57th Street, New York, New York 10019

Sports Afield, 250 West 55th Avenue, New York City, New York 10019

When the magazines arrive, study the classified advertising in their back pages. Do not get discouraged over the large number of ads you will see under the "live bait" heading because all of the ads would not be there if they did not sell. As a matter of fact, this is the marketplace for worm growers, bait dealers, organic gardeners and farmers, and fishermen.

Also consider advertising in both local and national newspapers. Your local library will have a copy of the *Ayers Newspaper Directory* which lists the names and addresses of thousands of newspapers across the United States.

Now send some more letters to several of the worm growers listed in the classified section, and the more the better. Tell them that you are interested in buying worms and would like to have complete price information and instructions on caring for worms. When you read their replies, you will gain considerable insight on how other successful growers operate.

Write to the bait container manufacturers listed below and explain that you will soon be shipping earthworms. Ask for price and descriptive information and any other literature they have available about their packing and shipping recommendations. Even though there will probably be a small charge for samples, I recommend you purchase them.

Keiding, Inc., 4545 West Woolworth Avenue, Milwaukee, Wisconsin 53218

Sealright Co., Inc., 4209 Noakes, Los Angeles, California 90023

Trico Container Corp., 5710 East Shiela, City of Commerce, California 90040   726-1121

Before placing your ads, spend some time planning your entire project from beginning to end, keeping it simple. Do not complicate your life worrying about how "big business" would go about planning with attorneys, accountants, and computers. This is a simple business. If you have questions about how to best plan your project, there are hundreds of competent people ready to help you free of charge. Visit your post office, stationery store, small business administration, chamber of commerce, better business bureau, city hall, and, most certainly, your banker. They will be happy to help you and will provide you with more printed information than you will have time to read. Also look into regulations concerning a business license and state and federal taxes.

Part of your planning should include an actual trial run of handling an order, including harvesting, packaging, and shipping your worms. For example, pretend that you have received an order for 200 worms. Make the effort to harvest the worms, package them, and then mail them to a friend or relative living some distance away who is willing to experiment with you. When the package arrives, have your friend inspect the container and count the number of worms. You may discover that most of the worms are dead or the container is inadequate or the shipment was too long in transit. It is best to solve these kinds of problems before you place your ads.

After your ads appear in the magazines, you will not only receive orders but hundreds of inquiries as well. As a matter of fact, your ads should probably appeal for inquiries because it is simply too costly to tell the whole story in advertisements of this kind. I also believe it is a mistake to show price information in your ads. In any event, you will have to be prepared to reply to inquiries, which represent potential orders, in a persuasive, professional way; the best way to be prepared is to be more knowledgeable than any of your competitors. Force yourself to study, and if possible, go out and visit bait shops, dealers, and organic gardeners and farmers. Ask all the questions you can think of and read everything available on the earthworm industry. Write to the U.S. Department of Commerce, Washington, D.C., and ask them for all available information on the mail-order business and the worm business. Also, in your letters to magazine publishers, ask them to send you information on how to write effective, order-pulling ads.

Remember the following:

Keep your business simple.
Ask questions.
Follow instructions.
Be prepared.
Continually add to your knowledge.

If you follow these basic steps, you will succeed in selling worms.

Other Ways To Sell Worms

You purchase most of the products and services you need for your family, such as groceries, clothing, hardware, and laundry, from retail stores or so-called discount outlets. You purchase from stores or service centers located near where you live or work, both as a matter of convenience and to conserve gasoline and driving time. This helps explain why almost every community in the United States is self-contained with its own cluster of shopping and service centers. When new major tracts of homes or apartments are developed, new shopping centers invariably follow.

In relation to your selling earthworms and castings, this means that every shopping center is a source of steady, cash-paying customers, whose only purpose in being there is to buy. Take advantage of this dynamic marketplace; millions of dollars are spent to create shopping centers and to attract and keep customers coming back.

Dress up and call on the stores that might handle the sale of worms and castings. Agree to consign your worm and casting packages with the understanding that you will remove any damaged containers. It is important that you provide complete written instructions on the care and use of your products. Describe how nature created the earthworm to improve soil. All of your posters, packages, instructions, literature, and labels should be attractive and preferably designed by someone talented in color typography and illustration.

Offer the store owner a 40 percent discount from your asking retail price, and agree to do anything reasonable to assist the manager in selling your merchandise. Earthworm castings have an indefinite shelf life, but worm cartons should be rotated about once every two weeks. Remove the old, unsold cartons and replace them with new, fresh products. Unsold cartons can be conveniently emptied back into your beds for future use.

In every large shopping center, you will usually find one or more of the major retail stores. Do not be reluctant to call on them simply because of their huge size. With the right approach, you might be permitted to operate from a display booth in one of these big stores. If you plan your project right, you will sell more worms and castings than you can imagine. One tip to follow: always get the name and address of your customer or prospect, for both follow-up and repeat sales.

Let your imagination work for you and take advantage of shopping centers as your marketplace for worms and castings. For example, every shopping center has at least one or more bulletin boards for the use and convenience of the general public. These are usually located in grocery markets and laundromats, and they are used to advertise such things as baby-sitting services, used cars, and rewards for lost dogs. Take advantage of this free advertising space because your message will be read by hundreds of prospects every day. Again, your message should be clear, easy to read, and designed to catch the eye with colorful artwork.

Another way of getting results is through the free newspapers, printed locally and published cooperatively by the store owners. These normally offer either free or inexpensive advertising.

Helping Others Get into the Worm Business

Competition is both creative and productive. Our economic system is based on encouraging competition and free enterprise, which is regulated by the federal antitrust laws and assisted by the small business administration.

You have a tremendous financial opportunity awaiting you by helping others into the earthworm business. There are two important facts to consider. First, the earthworm business is on the verge of emerging into an agricultural giant. Second, try to understand that the law of supply and demand works contrary to what most of us believe about it. Simply put, supply always precedes demand. Automobiles, airplanes, garbage disposals, television sets, and tape recorders were all nonexistent at one time. They were created at great financial risk and introduced before any demand took place. The record business is a prime example of where the product is supplied before any demand takes place; stars like John Denver, Elton John, and Barbra Streisand were completely unknown until their talent was supplied as a recording to the public. The current supply and production of earthworms are not adequate to satisfy the current demand. Therefore, the whole worm business could fizzle—no earthworms, no demand!

Thousands of new growers will be needed to meet the demand for earthworms. Many of the newcomers will fail, not through lack of opportunity, but through loss of interest or inability to run a business. They will sell their farms or merge them into larger, more successful operations. In time, cooperatives will be formed to handle worldwide marketing and distribution. It will become big business, and the best part is that you are on the ground floor of this evolving business. You can profit by helping others into the earthworm business.

Where are the prospective growers? Where can you find them? The answers to these questions will be obvious once you understand the principles behind one of the biggest hoaxes perpetrated on the American public—the money-making-scheme segment of the mail-order business.

On the whole, the mail-order industry is legitimate and plays an important role in our lives. The majority of mail-order firms, including such giants as Sears, Montgomery Ward, Walter Drake, and Sunset House, do provide worthwhile service to their mail-order customers. Their products are excellent and always backed up with money-back guarantees. Generally you cannot go wrong making purchases through the mail unless you buy money-making schemes.

The opportunity seeker offers you one of the best avenues for successful marketing in the earthworm business. Year after year, millions of people take a plunge into the mail-order business—a false and costly plunge. In a way, they are innocent victims of greed. There are no easy roads to riches; but people still follow the wrong road leading nowhere, blindly throwing away their hard-earned money on money-making schemes.

Millions of opportunity seekers are waiting for you to set them up in a legitimate, worthwhile business that does not require a lot of time or money. If these same people are willing to spend valuable time and money chasing rainbows, think of what you have to offer them in the earthworm opportunity. The remaining question is how to reach your prospects.

To reach opportunity seekers, do exactly as they do: reply to all the money-making ads you can find in magazines sold in grocery and liquor stores. The ads appear in the classified advertising section under listings titled such as "business opportunity," "money-making ideas," and "of interest to women." In a few days, you will be overwhelmed with junk mail that will all pitch the same theme: send money and we will make you rich.

Purchase every offer you can afford; they will range from $1 to $10. Every dollar you spend on this project will be worth its weight in gold. You will be buying courses or instructions teaching you how to sell mail order. Notice that I did not say "how to sell *by* mail order." Study the material carefully and completely to learn how to get free advertising, free lists of opportunity-seeker names, names of circular mailers and commission mailers, and information on low-cost cooperative advertising. This will be your introduction to a gigantic business enterprise. You will get a sound education in the business-opportunity segment of the mail-order business. If you take advantage of everything you learn and put it into practice, you will make a fortune selling worms.

Growing Worms in an Apartment

Growing earthworms for a living is anything but dull. In my daily business and social contacts, I find it rewarding to answer questions about a subject that creates so much interest. Everyone who learns about earthworms seems to hunger for more information.

When you tell people about earthworms, the reaction is unreal. Some of your friends and neighbors will chuckle while others will think you have lost touch with the real world, but you will get a positive reaction and be flooded with questions like, "I beg your pardon? What did you say? Repeat what you just said!"

When I first started growing worms, I was very reluctant to mention my project because it seemed embarrassing and a kind of lowly thing to be doing. That is, until I met Toni. I was first introduced to Toni at one of those polite cocktail parties where everyone wears a suit and tie or a long dress. I took an immediate liking to her; she was attractive and personable. All during the evening I wanted to mention my worm project, but decided that the risk of offending her was too high.

Cocktail parties have a way of changing attitudes; if you are like me, sooner or later you say anything! So it was that I told Toni about my worms. Her reaction was totally unexpected: "Well, can you believe that! Hank, I grew earthworms in my apartment over 30 years ago."

In 1945, Toni was a single stewardess for a major airline. In those days her salary was $125 a month, less the cost and maintenance of her uniforms. On that kind of money, survival was a real problem. Fortunately, her parents owned a small apartment house that Toni and her good friend Suzy shared and managed on a rent-free arrangement. Suzy was also an airline stewardess suffering from the same shortage of funds. One of their favorite pastimes was asking, "What can we do to make more money?"

Before leaving on a vacation in her hometown, Suzy told Toni about her aunt who made money raising worms for fishermen. Could they do the same thing to make extra money and still keep their jobs? They chose the flower box to house the worms. Two weeks later, Suzy returned with a box full of squirming worms and some handwritten instructions.

The flower box was located on the back side of their kitchen, running the entire 10-foot length. The box was 2-feet across and about 12-inches deep. They removed all of the dirt and the flowers and plants that were happily growing there. They then filled the empty planter with various organic wastes collected from the neighborhood, including sawdust, leaves, grass clippings, manure, and garbage. As instructed, they sprinkled on limestone flour, thoroughly wet the contents, and added the worms. They were now in the worm business.

The worms grew and multiplied; in time, more space was needed to house the growing population so Toni and Suzy expanded their operation into the flower beds surrounding their small patio, replacing the plants and dirt with more grass clippings, leaves, manure, and garbage.

The girls had fun with their enterprise, conserving every nickel and dime to help their business succeed. The small apartment soon became a factory. Cans were stacked everywhere, and Toni and Suzy were busy counting and packing worms. On their scheduled flights, they would smuggle the cans aboard the airplane and deliver their worms in cities across the country. They sold the worms at a penny a piece. The business thrived and Toni never lost interest in the earthworm opportunity—she is currently a partner in one of the largest worm ranches in California.

When To Begin Selling Worms

Your decision on when to begin selling worms will eventually have to be made. This is an economic question; remember, the longer you wait, the more worms you will have to sell. Every worm you sell will be gone forever and the reproductive possibility lost. Refer back to page 32 and study that section again. Look what happened to the original 100,000 worm population after 2½ years.

Do you have any notion about the value of those worms? Would they be worth $50,000 or $100,000? No, we are way off. The total worm population 2½ years later at today's average retail price would be worth almost $1,500,000!

I am not suggesting you wait 2½ years before you start selling your worms; that is a long time to wait, and housing that many worms would require a great number of beds plus the real estate to put them on. However, if your goal is to make a large, lump-sum of money, then all you have to do is keep growing and splitting your beds for a long time. If your goal is to remain small and earn a steady monthly income, just decide on the number of beds you are willing to maintain.

The standard 4-foot-by-8-foot bed will produce about ten pounds of worms a month. The average retail price, as of this writing, is $8 per pound for bed-run worms. Ten beds will produce a monthly, year-round gross income of $800; 20 beds will double this figure. Keep in mind that your castings will produce additional annual income.

Cost of Getting Started in Worm Ranching

The only real and significant cost in starting your worm business will be the cost of your initial worms. How much or how little you spend is entirely your decision; but the more you buy, the sooner you will reach your goal. You should ask yourself these questions: How much can I afford? How long can I wait before selling my worms?

The cost of lumber, nails, hinges, limestone, tools, water, and electricity are all low-cost items and easy to figure. Approximating necessary land-space cost may be more difficult.

Based on personal experience, I believe that once you take the initial step, the question of space will solve itself. I solved the problem by driving around looking for unused land behind homes and ranches and offering a small rental fee for its use. I offered a fee of $1 per month per bed, including electricity and water. If you follow this plan, numbers of people will be interested in what you are doing and will offer you all the rental land you need. My entire ranch is located on rented ground at the rate of $1 per month per bed. After running out of room in his backyard, another grower solved the problem by renting the backyards of four adjoining neighbors' properties. Still another grower, with extremely limited space, constructed his beds bunk-bed style, building one bed on top of the other.

The cost of starting your operation is an investment that will surely grow if you put your mind and energy to your project. Remember, if you start with just one bed of worms and have the willingness and patience to wait one year before beginning to sell your worms, you will enjoy an annual income surpassing that of many entertainment and sports personalities and hard-working, worn-out executives.

Now that is some investment!

Bunk-bed style beds designed for limited space in the backyard.

Raising African Night Crawlers

The African Night Crawler is the largest of all earthworm species and can grow to a length of 12 inches or more. Fishermen seem to prefer these giant worms because of their long, thick bodies which can be cut into several pieces of bait. Night Crawlers are in demand by fishermen and are sold at premium prices two to three times the price of an equal count of Red Worms. If you choose to advertise and sell earthworms to fishermen, you should consider raising this species because they could add tremendously to your income; but consider the problems.

The chief problem is temperature. African Night Crawlers are found growing naturally only in countries with tropical climates. They cannot survive temperatures below 50 degrees, and their environment or bedding temperature must be controlled in the 70 to 80 degree range. To successfully grow Night Crawlers requires indoor housing with sophisticated heating and temperature-control equipment.

Just as their name implies, African Night Crawlers do indeed crawl. The use of light is the surest way to keep earthworms in their bed because they retreat from light. Be prepared to hang lights over the beds, preferably with an automatic timer adjusted for nightfall.

African Night Crawlers are commercially grown and sold all across the United States, even in the cold regions. I do not grow Night Crawlers and, therefore, am not an authority on them. I suggest that you reply to ads in which Night Crawlers are specifically advertised. The growers will send you complete and detailed information on the care and marketability of these huge, profitable earthworms.

Conclusion

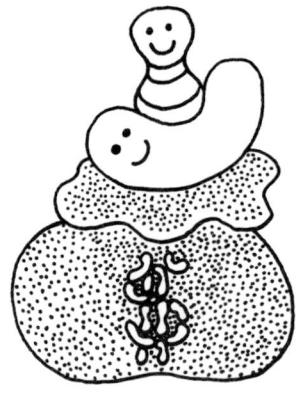

You should now go back to the beginning and review all of the material in this book. The information is valuable, but it is not entirely complete. The only way for you to get the complete picture of the worm business is to begin in it.

If you do take positive action, before long you will experience the confidence and ability to operate successfully. You will even find yourself experimenting with such things as feeds, bed construction, watering schedules, harvesting methods, packaging, and advertising. Your discoveries will often disagree with what you have read, but your ideas may turn out to be better. Self-determination is good and surely part of the excitement of operating your own worm business.

As you review the material, jot down questions and make notes on information that seems unclear. This will aid you in your study. Before commencing your review, let us go over a few last-minute questions.

Can I start in the worm business by digging up worms out of my garden?

No. Garden worms are not commercially grown because they reproduce too slowly.

What is the life span of Red Worms?

Ten to 15 years.

Can worms be grown in cold climates?

Yes. There are hundreds of growers located in the colder regions of the United States and in Canada.

How do I protect the worms from freezing temperatures?

Place a heavy layer of feed, about 6 inches in depth, over the entire bed surface. Then put hay or straw along the edges and on top of the feed layer, using as much hay or straw as you can conveniently get into the bed.

Will the worms crawl into the earth below the bed?

If the bedding is kept moist and supplied with feed, the worms will not crawl into the earth below the bed.

Have any special breeds of Red Worms been developed?

No. There are several growers who advertise their worms as having special characteristics, but it is not true; Red Worms are all the same (lumbricus rubellus).

How many worms weigh one pound?

It depends on the size and maturity of the worms. Generally, either 1,000 mature worms, 2,000 medium-size worms, or 2,500 mixed-size worms (bed run) will weigh about one pound.

Why should I keep dirt out of the bed?

It is not essential to keep dirt out of the bed, but dirt does tend to compact so the bedding may require more frequent turning to keep it loose.

Is odor a problem in worm growing?

No. Regular use of limestone flour will eliminate all disagreeable odors in both worm beds and stored manure.

How many beds of worms can I personally manage without the use of outside help?

You can easily maintain 100 beds, including everything from watering, feeding, harvesting, packing, and shipping to handling all of the office work, with plenty of time off for leisure.

What size container should I build for each pound of bed-run worms (2,500 worms)?

Use 1 cubic foot for each pound of bed-run worms. You can easily estimate the cubic feet of a bed by multiplying the width by the length by the height. A container measuring 2 feet by 2 feet by 1 foot equals 4 cubic feet. A 4-cubic-foot container will hold 10,000 worms, that is, four pounds, easily.

When I pack worms for shipment, how much peat moss should I use?

First of all, use only Canadian peat moss. Soak it for at least 24 hours, and then remove the excess moisture by squeezing the peat by hand. The moisture content will be about right when you can squeeze out only about 2 to 3 drops of water. In your shipping containers, use about ¾ volume of peat moss to ¼ volume of worms. Store all containers in a cool, dark area prior to shipping.

Can I really harvest ten pounds per month from each of my beds?

It depends on the size of the bed and the number of worms occupying it. If your bed has only 10,000 worms in it, you obviously are not going to harvest ten pounds. If you expect maximum production, your beds should contain a worm population of about 200,000 before you do any harvesting. Under proper care and normal conditions, this population should produce ten pounds per month.

How can I determine the exact number of worms in my beds?

This may seem troublesome, but I suggest you harvest one complete bed and weigh the worms. You should do this frequently in the beginning to give you a feel for worm population at various stages of growth.

Do worms lose weight?

Yes. They will lose both weight and size if not properly fed and watered. Also, if the beds are not properly maintained, the survival rate of babies and capsules will be poor.

How many worms should I harvest from my beds at one time?

The rule of thumb is not to remove more than one-third of the worms. In theory, the remaining population of two-thirds will make up for the loss in about 30 days.

What would be wrong with harvesting all of the worms at one time?

Nothing. Many ranchers harvest all of the worms except for babies and capsules and then wait a few months for the remaining worms to multiply.

Can I use hay or alfalfa as feed for my worms?

Yes except that hay decomposes very slowly and does not absorb moisture very well. It can be successfully used if it is mixed with manure and reduced in particle size by a shredding machine.

Do you have any program to help me get started in the worm business?

This book is your program. Drop me a line if you would like price information on earthworms that I sell.

Will the earthworm business become saturated with growers?

Yes, but the hamburger business was saturated when McDonald's started out, too.

Are there any worm ranchers now who cannot sell their worms?

Yes. Plenty of them! I see their ads in newspapers advertising full beds of worms for as little as $50 per bed. This is ridiculous, yet understandable because some people run scared and others lack the energy to succeed. If they would follow the instructions outlined in this book, they would not have trouble selling worms.

What is a buy-back agreement?

Some of the early promoters in the earthworm business agreed to buy back all of the worms you could ship them at the going wholesale price. The catch was that you had to buy your initial beds from them. Their loose agreements called for some clever restrictions such as the size, health, and condition of the worms you shipped to them. If you are ever offered this kind of arrangement, beware and consult your attorney before you part with your hard-earned money.

Has any attempt been made to organize earthworm growers?

The only organization that I can recommend your joining at this time is the San Diego County Farm Bureau. This is a nonprofit, tax-exempt organization that is part of the national Farm Bureau network with over 2½ million members. They have just established an earthworm commodity division, in cooperation with the U.S. Department of Agriculture, to aid in the research, development, and marketing of earthworms. For information write them at the San Diego Farm Bureau, 1670 East Valley Parkway, Escondido, California 92027.

In conclusion, let me make one final point: the reason you have such an exceptional opportunity in the worm business is that the industry is still in its infancy. Much like a rocket about to be launched, once it takes off it will not slow down for years to come, if ever.

When I roll out of bed in the morning, my first thoughts are so darned pleasant it almost scares me. All I can think of is worms! That probably sounds silly, but the little creatures have been good to me—just as they will be good to you, too!

I no longer have to struggle with long hours at my job. I can go fishing or hunting when I decide to and stay as long as I desire. Being independent, I can choose how much money I make.

I wrote this book to help you get started on the same road to happiness. Thousands of people across the land, from all walks of life, have read this book; I receive letters from widowers, firemen, students, retirees, airline stewardesses, teachers, doctors, salesmen, and those out of work. It is quite a feeling to help people find opportunity.

Today is the day for you to get started. Do it today and end up like me: happy, easy going, and rich!

Good luck and good worming.